This Is a Long Journey of my understanding the world of Quantum Realm and this is just the beginning of new and great series

Index

HOW DOES QUANTUM ENTANGLEMENT EFFECT PARTICLES

Before we dive into Quantum entanglement we will talk about Quantum Mechanics.

Now what is Quantum Mechanics?

Quantum Mechanics is the study of tiny lengths. When we talk about tiny lengths we talk about particles like: sub atomic particles, photons etc. Now let us say that we only talk about particle nature.

The nature of the particle exists in two different nature, one is wave nature and the other is particle nature. And these two natures are separated by something called as a measurement barrier. Measurement barrier is the measurement of the wave function which collapses and give use the results in particles. The only difference is that the our knowledge has a huge gap on how the wave function collapses. This concept is called the **Particle Duality**.

Now let's see what we get to know about the famous double slit experiment.

Double Slit Experiment.

The double slit experiment is a famous experiment in physics that the demonstrates the wave-particle duality of light and matter. It was first performed by Thomas Young in early 1800s and has since been replicated with various particles, including electrons and even large molecules.

In the basic setup of the double slit experiment, a coherent light source, such as laser, is directed towards a barrier that contains two small slits close together as sown in the figure below.

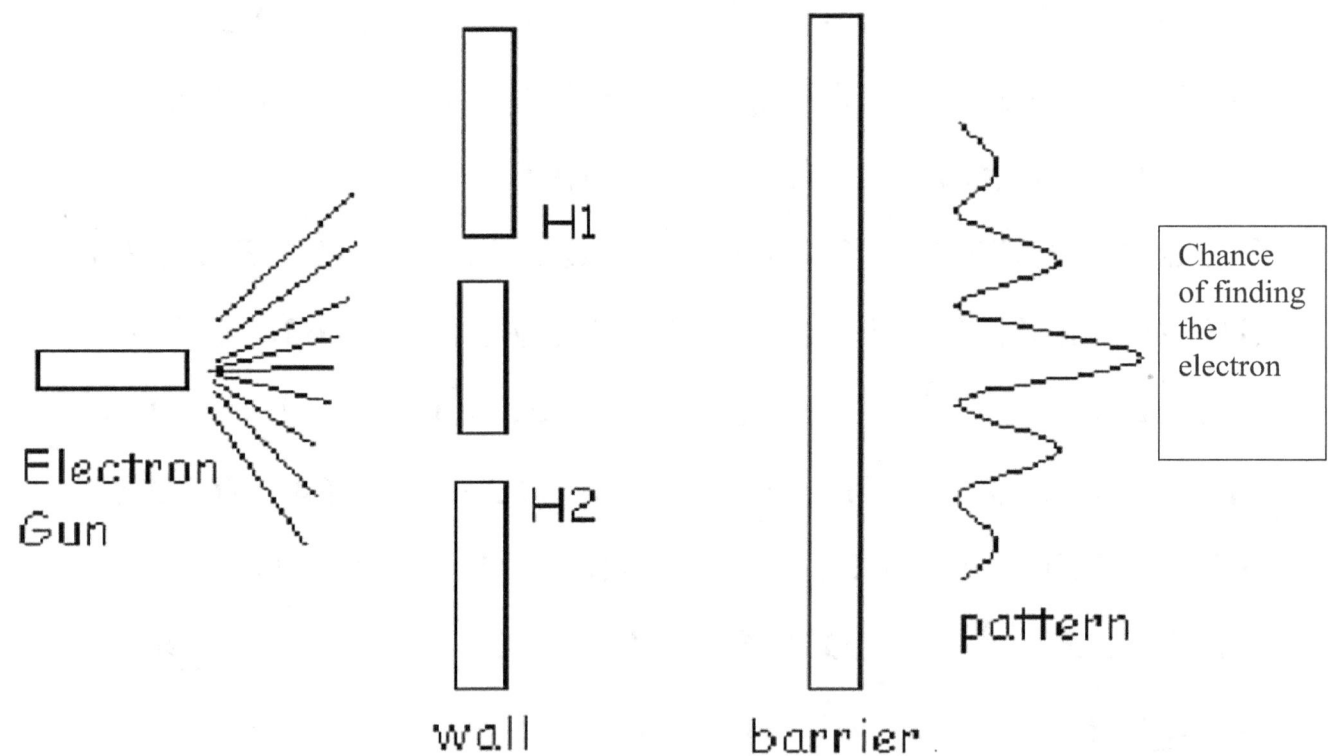

Behind the barrier, there is a screen or a photographic plate that captures the pattern produced by the light passing through the slits. When light passes through the slits, it distracts, or spreads out, and creates an interference pattern on the screen. This pattern consists of alternating bright and dark fringes, resulting from the constructive and destructive interference of the light waves.

The surprising part of the experiments comes to the intensity of light is reduced so much that it is emitted as 1 photon at a time. Despite sending individual photons over time. The pattern that emerges on the screen is still an interference pattern. This suggests that each photon behaves as if it passes through both slits and interferes with itself, even though it's a particle.

The phenomenon can be explained by the wave-particle duality principle according to which, *__"PARTICLES SUCH AS PHOTONS OR ELECTRONS CAN EXHIBIT BOTH WAVE AND PARTICLE LIKE PROPERTIES DEPENDING ON HOW THEY'RE__*

MEASURED OR OBSERVED". In the case of the double-slit experiment, the photons or electrons can be thought of as waves which interfere with each other, thus cresting the observed pattern.

The double-slit experiment has certain profound implication for the understanding of quantum mechanics and nature of reality. It demonstrates that at the quantum level, the particles do not possess definite properties until they're measured or observed. The experiment also highlights the importance of the observer's role and the delicate relationship between the observer and the observed system.
The double-slit experiment continues to be an on-going research and has paved the way for further exploration into the fundamental nature of the particles and the wave particle duality. It remains a corner stone of quantum physics and has inspired numerous discussions and interpretations in the scientific community.

Wave Particle Duality

Wave-particle duality is a fundamental concept in quantum mechanics that describes the dual nature of particles. It suggests that particles, such as photons or electrons, can exhibit both wave-like and particle-like properties, depending on the experimental setup and how they are observed.

On one hand, particles exhibit characteristics commonly associated with particles. They have mass, occupy a particular position in space, and can be localized. This particle nature is often described by their discrete energy levels and the ability to interact as individual entities.

On the other hand, particles also display wave-like properties. They can exhibit interference and diffraction patterns, similar to what is observed with waves. These characteristics are typically associated with the behaviour of waves, such as light waves or sound waves. Waves have properties like wavelength, frequency, and amplitude, and they can propagate and interfere with each other.

The wave-particle duality principle suggests that particles, at the quantum level, possess both particle-like and wave-like properties simultaneously. This means that a particle, such as an electron, can exhibit wave-like behaviour, such as interference or diffraction, under certain experimental conditions. However, when measured or observed, the particle appears to "collapse" into a localized state, behaving like a particle with specific properties, such as position or momentum.

The behaviour of particles as waves is described by wave functions in quantum mechanics. These wave functions represent the probability distribution of finding a particle in a particular state. They can be used to calculate the likelihood of a particle being found in a specific position or having a certain energy. The wave function evolves over time according to the Schrödinger equation, which describes the dynamics of quantum systems.

The wave-particle duality principle has been experimentally confirmed through various experiments, including the double-slit experiment mentioned earlier. These experiments demonstrate that particles, when not observed, can exhibit wave-like behaviour and interfere with themselves. However, when observed, they behave as localized particles with definite properties.

Wave-particle duality challenges our classical intuitions about the nature of particles and waves. It suggests that the behaviour of particles is inherently probabilistic and that the act of observation or measurement influences the outcome. This concept is at the heart of quantum mechanics and has significant implications for our understanding of the microscopic world.

Now What Is Planck's Theory?

Planck's theory also known as Planck's quantum theory refers to the ground breaking work of the German physicist Max Planck in the early 20th century. Planck's revolutionized our understanding of the behaviour of the development of quantum mechanics.

In 1900, Planck was studying the problem of black body radiation, which is the electromagnetic radiation emitted by the object at a given temperature. Classical physics predicted that the amount of radiation emitted at each wavelength should increase indefinitely as the wavelength decreases, leading to what are known as "ultraviolet catastrophe". However, experimental observations did not match these predictions.

To resolve this discrepancy, Planck's proposed a radical idea. He suggested that energy of electromagnetic radiation is not continuous but is quantized, meaning it can only take certain discrete values. Planck postulated that energy is emitted or absorbed in tiny discrete packets, which is called "quanta" (singular: quantum). The energy of each proportional to its frequency, as described by the equation $E = hf$, where E is the energy, h is the Planck's constant (a fundamental constant of nature), and f is the frequency of the radiation.

This concept was a departure from classical physics, which assumed that energy could be infinitely subdivided and that the energy of a system could have any value. Planck's theory introduced the idea that energy is quantized and that the energy of a system can only change in discrete amounts.

Planck's quantum theory successfully explained the observed behaviour of black-body radiation, matching the experimental data. It provided a theoretical framework that could account for the observed discrete nature of energy in certain physical systems.

Planck's work had profound implications for the development of quantum mechanics, as it paved the way for further investigations into the quantized behaviour of particles and electromagnetic radiation. His theory laid the foundation for the later developments by physicists like Albert Einstein, Niels Bohr, and others, leading to the formulation of a more comprehensive quantum theory that revolutionized our understanding of the microscopic world.

Planck's constant, h, remains a fundamental constant in quantum mechanics and is used in various equations to describe the behaviour of the particles and waves on the quantum scale. It is a fundamental constant of nature, determining the relationship between energy and frequency in the quantum realm.

What is ultraviolet catastrophe?

The ultraviolet catastrophe refers to a theoretical problem in classical physics that arose in the late 19th century when attempting to describe the distribution of energy in black-body radiation. It was a major puzzle that remained unresolved until Max Planck proposed his quantum theory in 1900.

A black body is an idealized object that absorbs all incident electromagnetic radiation, regardless of frequency or wavelength. When a black body is heated, it emits radiation across a wide range of frequencies, including visible light. Classical physics predicted that the amount of radiation emitted at each wavelength would increase without limit as the wavelength decreased. This implied that an infinite amount of energy would be emitted in the ultraviolet region of the spectrum, resulting in an infinite total energy.

This prediction contradicted experimental observations. Experiments had shown that the energy distribution of black-body radiation peaked at shorter wavelengths, but then decreased as the wavelength approached the ultraviolet region. This implied a finite amount of energy being emitted, which was inconsistent with classical predictions.

The inability of classical physics to explain this contradiction between theory and experiment became known as the ultraviolet catastrophe. It highlighted a fundamental flaw in classical electromagnetic theory and raised questions about the nature of energy and radiation.

Max Planck proposed a solution to this problem by introducing the concept of energy quantization, which led to the development of his quantum theory. Planck postulated that energy is emitted or absorbed in discrete packets, or quanta, rather than being continuously distributed. This allowed him to derive an equation that matched the observed energy distribution of black-body radiation, resolving the ultraviolet catastrophe.

Planck's quantum theory not only solved the problem of black-body radiation but also laid the foundation for the development of quantum mechanics. It introduced the idea that energy is quantized and can only change in discrete amounts, revolutionizing our understanding of the behaviour of particles and electromagnetic radiation at the microscopic level.

What is Super Position

Superposition is a fundamental concept in quantum mechanics that describes a property of quantum system, such as particles or qubits, where they can exist in multiple states simultaneously. In other words, instead of being confined to single well-defined state like on classical physics, a quantum system in a state of superposition can be in a combination or "superposition"

Of multiple states.

In classical physics, we are accustomed to thinking of objects as being in a well-defined state. For example, a light well defined state switch can be either on or off, a coin can be heads or tails, and a cat can be alive or dead. However, in the quantum realm, particles can be state that is a combination of two or more distinct states.

Mathematically, superposition is represented using a linear combination of basis states. For example, in the case of a qubit, the two basis states are usually denoted as $|0\rangle$ and $|1\rangle$, which represent the classical states of 0 and 1. A qubit in a state of superposition can be described by the mathematical expression $\alpha|0\rangle + \beta|1\rangle$, where α and β are complex numbers called probability amplitudes. The square of the magnitude of this probability of measuring the qubit in the corresponding basis states.

The most famous example illustrating superposition is the thought experiment known as Schrödinger's cat. In this scenario, a cat is imagined to be in a superposition of being both alive and dead until it is observed or measured, at which point it collapses into one of the states. This thought experiment highlights the peculiar nature of quantum superposition and the role of observation or measurement in determining the outcome.

Superposition is a crucial property in quantum computing. By exploiting the superposition of qubits, quantum computers can perform computations on multiple states simultaneously, offering the potential for exponentially faster processing for certain problems compared to classical computers.

It's important to note that superposition is a fundamental aspect of quantum mechanics and distinguishes it from classical physics. While superposition may seem counter-intuitive from our everyday experience, it has been experimentally confirmed through various quantum interference experiments and is a cornerstone of quantum theory.

Let Us Know What Quantum Entanglement Is:

Now let us take two particles where they are some how entangled with respect to infinite space, time and distance.

When the particles are said to be entangled they are information correlates to each other.

The effect of quantum entanglements on particles is that it establishes a strong correlation between their properties. For example, let's consider a pair of entangled particles with spin, a property associated with

fundamental particles. If the spins of the entangled particles are in a super position, meaning they can be both up and down simultaneously, the entangled particles will exhibit a correlated behaviour.

Let's say that we measure one particle's spin number and it turns out to be positive 1/2 then automatically the other particles spin number would be negative 1/2. We can say that the outcomes of measuring the first particle would result the information of the second particle would to be the exact opposite of the first particle.

When a measurement is performed on one of the particles, such as measuring its spin, the act of measurement "collapses" the superposition of the entangled particles into a definite state. Remarkably, the state of the other entangled particle instantaneously collapses as well, even if it is located at a great distance. This instantaneous correlation, known as "spooky action at a distance," puzzled many physicists, including Einstein, who referred to it as "spooky" due to its non-local nature.

Quantum entanglement is a fundamental concept in quantum mechanics and has been verified through numerous experiments. It has profound implications for our understanding of the nature of reality and has applications in various fields such as quantum computing, quantum cryptography, and quantum teleportation.

Quantum entanglement is a fundamental phenomenon in quantum mechanics where two or more particles become correlated in such a way that the state of one particle cannot be described independently of the other, regardless of the distance between them. Here are a few ideal examples of quantum entanglement:

1. **Bell State or EPR Pair**: This is one of the most famous examples of quantum entanglement. In an EPR (Einstein-Polyanovsky-Rosen) experiment, two particles, such as photons, are generated in a way that their quantum states become entangled. The particles are created in a

superposition of states, and measuring one particle's state instantaneously determines the state of the other particle, even if they are separated by large distances.

2. **Quantum Teleportation**: Quantum teleportation is a protocol that allows the transfer of quantum information from one location to another, without physically moving the particles themselves. It relies on the entanglement between two particles. By performing measurements on one particle and sending the results to the receiver, the state of a third particle can be recreated at a distant location.

3. **Quantum Cryptography**: Quantum entanglement plays a crucial role in quantum cryptography, which is a secure method of transmitting information. Quantum key distribution (QKD) protocols utilize the entanglement between particles to establish a shared secret key between two parties. The security of the key distribution process relies on

the principles of quantum mechanics and the impossibility of eavesdropping without being detected.

4. **Quantum Computing**: Quantum entanglement is a vital resource in quantum computing algorithms. Quantum bits or qubits can be entangled to perform computations that are not possible with classical bits. Entanglement allows for parallel processing and enables quantum computers to solve certain problems more efficiently than classical computers.

5. **Quantum Experiments and Tests of Foundations**: Quantum entanglement has been studied extensively to test the foundations of quantum mechanics and explore phenomena like non-locality and the violation of Bell inequalities. These experiments involve entangling particles and measuring their properties to investigate the nature of reality at the quantum level.

Let us understand few interesting topics related to quantum entanglement.

What is Bell's state or EPR PAIR?

Bell's state, also known as an EPR pair, refers to a specific entangled quantum state of two particles. It is named after physicist John Bell and refers to an aspect of the Einstein-Polyanovsky-Rosen (EPR) paradox.

The Bell state or EPR pair is typically described in the context of two spin-1/2 particles, such as electrons, which can be in one of two possible spin states: spin up or spin down. The state of the two particles is represented as:

$$|\Psi\rangle = (1/\sqrt{2})\,(|up\rangle \otimes |down\rangle - |down\rangle \otimes |up\rangle)$$

Here, |up⟩ and |down⟩ represent the spin-up and spin-down states of a particle, and ⊗ denotes the tensor product.

The important property of the Bell state is its entanglement. When the two particles are in a Bell state, their individual spins are undefined, but their spins are correlated in a specific way. If one particle is measured and found to be in the spin-up state, the other particle, when measured, will be in the spin-down state, and vice versa. This correlation is often referred to as "spooky action at a distance" because the measurement of one particle instantaneously affects the state of the other, regardless of the distance between them.

The Bell state has played a crucial role in tests of the foundations of quantum mechanics, particularly in experiments testing Bell's inequality. These experiments have shown that the predictions of quantum mechanics are inconsistent with classical theories that assume local realism, demonstrating the non-local nature of quantum entanglement.

Quantum Teleportation

Quantum teleportation is a remarkable phenomenon in quantum mechanics that allows for the transfer of quantum information from one location to another without physically moving the quantum state itself. Despite its name, quantum teleportation does not involve the transportation of matter or energy but rather relies on the principles of quantum entanglement and classical communication.

The process of quantum teleportation involves three main components: the sender (Alice), the receiver (Bob), and the quantum state to be teleported (the target state). Here is a step-by-step explanation of how quantum teleportation works:

1. Initialization: Alice and Bob start with a pair of entangled particles, usually referred to as the "Bell pair." Each particle in the Bell pair is a qubit that can exist in a superposition of states. The entanglement ensures that the states of the two particles are correlated, regardless of the distance between them.

2. Entanglement of the target state: Alice wants to teleport the quantum state of another particle (the target state) to Bob. She brings the target state in contact with one of the particles from the Bell pair, causing the target state and the particle to become entangled. This creates a three-particle entangled state.

3. Measurement and communication: Alice performs a joint measurement on her particle of the entangled trio and the target state. This

measurement extracts certain classical information about the quantum state of the target. The result of this measurement is a pair of classical bits.

4. Transmission of measurement results: Alice communicates the two classical bits of information (the measurement results) to Bob through a classical communication channel. This communication does not violate the principles of relativity since it involves sending classical information rather than quantum information.

5. Transformation of Bob's qubit: Based on the received classical information from Alice, Bob applies a specific set of quantum operations, known as a quantum gate, to his particle of the Bell pair. These operations depend on the measurement results and are designed to transform Bob's particle into an exact replica of the original target state.

At this point, Bob's particle now holds a quantum state that is identical to the original target state. The quantum information has effectively been teleported from Alice to Bob.

It's crucial to note that the process of quantum teleportation relies on the pre-existing entanglement between Alice's and Bob's particles, which needs to be established and maintained beforehand. The actual teleportation occurs when the quantum information is transferred by communicating the classical measurement results.

Quantum teleportation has been experimentally demonstrated and is an essential tool in various quantum information processing tasks. It plays a crucial role in quantum communication, quantum cryptography, and quantum computation, where the transfer of quantum states is necessary for performing complex computations or secures communication.

We can use these methods in improving technology in the form of qubits. And qubits can help developing new and improved method of computing called quantum computing

What are Qubits?

The correct term is "qubits," not "quantum bites." Let me explain qubits and their relation to quantum entanglement.

Qubits, short for quantum bits, are the fundamental units of information in quantum computing and quantum information theory. They are analogous to classical bits but differ significantly in their behaviour due to the principles of quantum mechanics.

While classical bits can represent either a 0 or a 1, qubits can exist in a superposition of both states simultaneously. This means that a qubit can be in a linear combination of the 0 and 1 states, represented mathematically as:

$$|\psi\rangle = \alpha|0\rangle + \beta|1\rangle,$$

where α and β are complex numbers called probability amplitudes. The probability of measuring a qubit in the state |0> is given by $|\alpha|^2$, and the probability of measuring it in the state |1> is $|\beta|^2$. Importantly, the sum of the squared magnitudes of the probability amplitudes must equal 1, ensuring that the qubit's total probability adds up to 1.

The ability of qubits to exist in superposition states allows for parallel processing and the potential for exponential computational power in quantum computing. By manipulating qubits and exploiting quantum gates, quantum algorithms can perform certain computations more efficiently than classical computers.

Now, let's discuss the relation between qubits and quantum entanglement. Quantum entanglement is a phenomenon where two or more qubits become correlated in such a way that the state of one qubit cannot be described independently of the others. When qubits are entangled, their states are intrinsically linked, even if they are separated by large distances.

Entangled qubits exhibit a property called entanglement superposition. This means that while each individual qubit can exist in a superposition of states, the overall state of the entangled qubits cannot be decomposed into independent states for each qubit. Instead, the combined state of the entangled qubits must be described as a whole.

For example, consider a pair of entangled qubits, often called a Bell pair or an EPR pair. The state of an EPR pair can be written as:

$$|\psi\rangle = (|00\rangle + |11\rangle) / \sqrt{2},$$

where |00> represents both qubits being in the state |0> and |11> represents both qubits being in the state |1>. In this entangled state, if one qubit is measured and found to be in the state |0>, the other qubit will instantaneously collapse into the state |0> as well, regardless of the spatial separation between them.

Quantum entanglement is a powerful resource in quantum information processing. It enables quantum teleportation, secures quantum communication through quantum cryptography, and enhances the computational capabilities of quantum algorithms.

In summary, qubits are the basic units of information in quantum computing, capable of existing in superposition states. Quantum entanglement describes the correlation between entangled qubits, where their states are intrinsically linked, leading to phenomena such as entanglement superposition and non-locality.

Quantum Computing

Quantum computing is a revolutionary approach to computation that leverages the principles of quantum mechanics, a branch of physics that describes the behaviour of matter and energy at the smallest scales. Traditional computers use bits, which can represent either a 0 or a 1, as the fundamental unit of information. In contrast, quantum computers use quantum bits, or qubits, which can represent a 0, a 1, or both simultaneously thanks to a phenomenon called superposition.

Superposition allows qubits to exist in a state that is a combination of 0 and 1, meaning that a qubit can be in multiple states simultaneously. This property of superposition is the key to the potential power of quantum computing. While a classical computer with n bits can represent only one of 2^n possible states at a given time, a quantum computer with n qubits can represent all 2^n possible states simultaneously.

Another crucial concept in quantum computing is entanglement. Entanglement is a phenomenon in which two or more qubits become correlated in such a way that the state of one qubit cannot be described independently of the others. When qubits are entangled, the state of one qubit can instantaneously affect the state of another, regardless of the physical distance between them. This property enables quantum computers to perform certain calculations much more efficiently than classical computers.

To carry out computations, quantum computers use quantum gates, which are the quantum equivalent of classical logic gates. Quantum gates manipulate the quantum states of qubits, allowing for operations such as superposition, entanglement, and interference. By applying a series of quantum gates to a set of qubits, quantum algorithms can be designed to solve specific problems more efficiently than classical algorithms.

One of the most famous algorithms in quantum computing is Shor's algorithm, which can factor large numbers exponentially faster than the best-known classical algorithms. This has significant implications for cryptography since many encryption schemes rely on the difficulty of factoring large numbers. Another well-known algorithm is Grover's algorithm, which can search an unsorted database quadratically faster than classical algorithms.

However, despite the potential advantages of quantum computing, building practical quantum computers faces several challenges. One major challenge is maintaining the delicate quantum states of qubits, as they are prone to errors due to environmental noise. Researchers are actively working on developing error-correction techniques to mitigate these errors.

Various physical systems are being explored as platforms for implementing qubits, including superconducting circuits, trapped ions, topological states of matter, and photonics. Each platform has its own advantages and challenges, and progress is being made to increase the number of qubits, improve qubit coherence times, and reduce error rates.

To carry out computations, quantum computers use quantum gates, which are the quantum equivalent of classical logic gates. Quantum gates manipulate the quantum states of qubits, allowing for operations such as superposition, entanglement, and interference. By applying a series of quantum gates to a set of qubits, quantum algorithms can be designed to solve specific problems more efficiently than classical algorithms.

One of the most famous algorithms in quantum computing is Shor's algorithm, which can factor large numbers exponentially faster than the best-known classical algorithms. This has significant implications for cryptography since many encryption schemes rely on the difficulty of factoring large numbers. Another well-known algorithm is Grover's algorithm, which can search an unsorted database quadratically faster than classical algorithms.

However, despite the potential advantages of quantum computing, building practical quantum computers faces several challenges. One major challenge is maintaining the delicate quantum states of qubits, as they are prone to errors due to environmental noise. Researchers are actively working on developing error-correction techniques to mitigate these errors.

Various physical systems are being explored as platforms for implementing qubits, including superconducting circuits, trapped ions, topological states of matter, and photonics. Each platform has its own advantages and challenges, and progress is being made to increase the number of qubits, improve qubit coherence times, and reduce error rates.

In conclusion, quantum computing is a field that exploits the principles of quantum mechanics to create powerful computers that can perform certain computations much faster than classical computers. By harnessing the properties of superposition and entanglement, quantum computers have the potential to revolutionize areas such as cryptography, optimization, drug discovery, and simulation of quantum systems. However, there are still significant technical challenges to overcome before practical quantum computers can be realized at scale.

Quantum entanglement theory predicts several possible outcomes and phenomena. Here are some key concepts related to quantum entanglement and their potential consequences:

1. Non-locality: Quantum entanglement is a phenomenon in which the quantum states of

two or more particles become intertwined in such a way that the state of one particle cannot be described independently of the others. This entanglement can persist even if the particles are separated by large distances. As a result, measuring the state of one entangled particle can instantaneously affect the state of its entangled partner, regardless of the physical separation between them. This non-local correlation is a fundamental aspect of quantum entanglement.

2. Spooky Action at a Distance: Albert Einstein famously referred to quantum entanglement as "spooky action at a distance" because the instantaneous correlation between entangled particles seems to violate the principles of locality and cause-and-effect relationships. However, these correlations do not enable faster-than-light communication or violate causality since they cannot be used to transmit information.

3. Bell's Theorem: Bell's theorem is a

mathematical proof that establishes the existence of certain correlations that cannot be explained by any local hidden variable theory. It provides a way to experimentally test whether two particles are entangled or if their correlations can be explained by classical means. Violations of Bell's inequalities provide strong evidence for the reality of quantum entanglement.

4. Quantum Teleportation: Quantum entanglement allows for a phenomenon known as quantum teleportation, which involves the transfer of quantum information from one location to another without physically moving the particle itself. By entangling two particles and performing certain measurements on one of them, the quantum state of a third particle (which is not entangled with the original particle) can be teleported to a distant location.

5. Quantum Key Non-locality is a concept in quantum mechanics that arises from the phenomenon of quantum entanglement. It

refers to the non-local correlations that exist between entangled particles, meaning that the state of one particle cannot be described independently of the others, even when the particles are physically separated by large distances.

The idea of non-locality is in contrast to classical physics, where the behaviour of a system is typically determined by local interactions between its individual components. In classical physics, the properties of an object can be understood by considering only its local interactions and the exchange of information through local signals travelling at or below the speed of light. However, quantum entanglement challenges this classical understanding.

When two particles become entangled, their quantum states become interconnected in a way that measuring one particle's state instantaneously affects the state of the other, regardless of the physical distance between them. This instantaneous influence between the entangled particles is what Einstein referred to as "spooky action at a distance."

The non-local correlations arising from entanglement have been experimentally confirmed through tests of Bell's theorem. Bell's theorem provides a mathematical framework to analyse the statistical correlations between measurements on entangled particles. It demonstrates that there exist certain correlations that cannot be explained by any local hidden variable theory, meaning that the correlations are inherently non-local in nature.

These experiments, such as the Bell test experiments, have consistently shown violations of Bell's inequalities, which strongly suggest the existence of non-local correlations and confirm the predictions of quantum mechanics. These violations imply that the entangled particles are instantaneously affecting each other's states, regardless of the separation between them.

It is important to note that while non-local

correlations exist between entangled particles, they do not allow for faster-than-light communication or violate causality. This is because the non-local correlations cannot be used to transmit information. The specific outcomes of measurements on one entangled particle are probabilistic, and no information can be transferred faster than the speed of light by manipulating the entangled states. The correlations can be observed only when comparing the statistical patterns of measurements made on a large number of entangled particles.

The phenomenon of non-locality challenges our classical intuition and requires us to rethink our understanding of how information and influence propagate in the universe. While we cannot explain the mechanism behind non-local correlations, quantum mechanics provides a mathematical framework that accurately describes and predicts the behaviour of entangled particles, and non-locality is a fundamental aspect of this theory. Distribution: Quantum entanglement plays a crucial role in quantum key distribution (QKD), a method for secure communication. QKD exploits the properties of entangled particles to establish a shared secret key between two parties. Any attempt to eavesdrop on the communication would disturb the entangled state, and this disturbance can be detected, ensuring the security of the key exchange.

6. Quantum Computing: Quantum entanglement is a fundamental resource in quantum computing. It enables the creation of quantum gates that exploit the superposition and interference of entangled qubits, allowing for parallel computation and potentially exponential speed-up for certain algorithms compared to classical computers.

These are some of the possible outcomes and implications of quantum entanglement theory. Quantum entanglement continues to be an active area of research, and its applications and implications are still being explored and understood.

Quantum teleportation is a remarkable phenomenon in quantum mechanics that allows for the transfer of quantum information from one location to another without physically moving the quantum state itself. Despite its name, quantum teleportation does not involve the transportation of matter or energy but rather relies on the principles of quantum entanglement and classical communication.

The process of quantum teleportation involves three main components: the sender (Alice), the receiver (Bob), and the quantum state to be teleported (the target state). Here is a step-by-step explanation of how quantum teleportation works:

1. Initialization: Alice and Bob start with a pair of entangled particles, usually referred to as the "Bell pair." Each particle in the Bell pair is a qubit that can exist in a superposition of states. The entanglement ensures that the states of the two particles are correlated, regardless of the distance between them.

2. Entanglement of the target state: Alice wants to teleport the quantum state of another particle (the target state) to Bob. She brings the target state in contact with one of the particles from the Bell pair, causing the target state and the particle to become entangled. This creates a three-particle entangled state.

3. Measurement and communication: Alice performs a joint measurement on her particle of the entangled trio and the target state. This measurement extracts certain classical information about the quantum state of the target. The result of this measurement is a pair of classical bits.

4. Transmission of measurement results: Alice communicates the two classical bits of information (the measurement results) to Bob through a classical communication channel. This communication does not violate the principles of relativity since it involves sending classical information rather than quantum information.

5. Transformation of Bob's qubit: Based on the received classical information from Alice, Bob applies a specific set of quantum operations, known as a quantum gate, to his particle of the Bell pair. These operations depend on the measurement results and are designed to transform Bob's particle into an exact replica of the original target state.

At this point, Bob's particle now holds a quantum state that is identical to the original target state. The quantum information has effectively been teleported from Alice to Bob.

It's crucial to note that the process of quantum teleportation relies on the pre-existing entanglement between Alice's and Bob's particles, which needs to be established and maintained beforehand. The actual teleportation occurs when the quantum information is transferred by communicating the classical measurement results.

Quantum teleportation has been experimentally demonstrated and is an essential tool in various quantum information processing tasks. It plays a crucial role in quantum communication, quantum cryptography, and quantum computation, where the transfer of quantum states is necessary for performing complex computations or secures communication.

Quantum entanglement has several potential applications in the field of technology. While some of these applications are still in the early stages of development, they hold promise for transforming various areas of technology. Here are some notable applications of quantum entanglement in the tech field:

1. Quantum Communication: Quantum entanglement can be used to create secure communication channels that are resistant to eavesdropping. Quantum Key Distribution (QKD) is a prime example. By exploiting the non-local correlations of entangled particles, QKD allows for the secure exchange of cryptographic keys, ensuring the confidentiality and integrity of

communication.

2. Quantum Cryptography: Quantum entanglement can enhance cryptographic techniques by providing secure encryption and authentication methods. For example, Quantum Key Distribution (QKD) protocols use entangled particles to distribute cryptographic keys securely, making it extremely difficult for adversaries to intercept or decode the information.

3. Quantum Computing: Quantum entanglement is a fundamental resource in quantum computing. It enables the creation of quantum gates and the manipulation of qubits to perform computations that are exponentially faster than classical counterparts. Entangled qubits allow for parallel computation, superposition, and interference, which are essential for executing quantum algorithms.

4. Quantum Sensing: Quantum entanglement can be harnessed to enhance the precision and

sensitivity of sensors. For instance, in quantum metrology, entangled states can be used to achieve higher accuracy in measurements of physical quantities such as time, distance, or magnetic fields. Quantum entanglement-based sensors have the potential to revolutionize fields like navigation, imaging, and environmental monitoring.

Quantum entanglement is a phenomenon in quantum physics where two or more particles become linked together in such a way that the state of one particle cannot be described independently of the other, even if they are far apart. Here are some key equations and theories related to quantum entanglement:

1. Bell's Theorem: Bell's theorem, formulated by physicist John Bell, is a mathematical proof that demonstrates the non-local nature of quantum entanglement. It shows that certain statistical predictions of quantum mechanics cannot be reproduced by any local hidden variable theory.

2. EPR Paradox: The EPR (Einstein-Podolsky-

Rosen) paradox is a thought experiment proposed by Albert Einstein, Boris Podolsky, and Nathan Rosen. It highlights the apparent conflict between the principles of quantum mechanics and the concept of local realism. The paradox involves entangled particles and suggests that measuring the state of one particle instantaneously affects the state of the other, regardless of the distance between them.

3. Schrödinger Equation: The Schrödinger equation is a fundamental equation in quantum mechanics that describes the time evolution of a quantum system. It is given by:

$$i\hbar\partial\psi/\partial t = \hat{H}\psi$$

Here, ψ represents the wave function of the system, t is time, $\hat{H}$ is the Hamiltonian operator representing the total energy of the system, i is the imaginary unit, and $\hbar$ is the reduced Planck's constant.

4. Bell Inequality: Bell's inequality is a mathematical inequality derived from the work

of John Bell. It provides a way to test whether a physical system obeys local realism or whether it exhibits the non-local correlations predicted by quantum mechanics. Violation of Bell's inequality indicates the presence of quantum entanglement.

5. Entanglement Entropy: Entanglement entropy is a measure of the amount of entanglement between different parts of a quantum system. It quantifies the degree of correlation between the states of the subsystems. One commonly used measure of entanglement entropy is the von Neumann entropy, given by:

$$S = -TR(\rho \log \rho)$$

Here, ρ is the density matrix of the system, and TR denotes the trace operation.

Uses of Quantum Entanglement:
Quantum entanglement has numerous applications in various fields. Here are a few notable ones:

1. **Quantum Teleportation:** Quantum entanglement enables the teleportation of the quantum state of a particle from one location to another without physically moving the particle itself. This has potential applications in secure communication and quantum computing.

2. **Quantum Cryptography:** Entanglement-based quantum cryptography schemes use the principles of quantum entanglement to secure communication channels. By leveraging the properties of entangled particles, it is possible to create unbreakable encryption keys and ensure secure transmission of information.

3. **Quantum Computing:** Quantum entanglement is a crucial resource for quantum computers. Entangled qubits can be used to perform parallel computations and enable faster processing than classical computers for certain tasks, such as factorization and optimization problems.

4. Quantum Sensors: Entanglement-enhanced sensors can offer improved precision in measuring various physical quantities, such as time, acceleration, and magnetic fields. Entanglement allows for higher sensitivity and resolution compared to classical sensors.

These are just a few examples of the equations and applications related to quantum entanglement. Quantum entanglement continues to be an active area of research and on-going studies may lead to new insights and applications in the future.

Entanglement entropy is a concept that arises in the field of quantum information theory and quantum many-body systems. It measures the degree of entanglement between different parts or subsystems of a quantum system.

In a quantum system composed of multiple particles or subsystems, the state of the entire system is described by a joint wave function. However, the interesting feature of entanglement is that it cannot be fully described by the individual wave functions of the subsystems. Instead, the entangled state requires a description in terms of the combined wave function of the entire system.

The entanglement entropy quantifies the entanglement between two or more subsystems by examining how information is distributed across them. It measures the amount of information or uncertainty associated with a particular subsystem when the other subsystems are ignored or traced out.

One commonly used measure of entanglement entropy is the von Neumann entropy, which is named after John von Neumann. For a bipartite system with subsystems A and B, the von Neumann entropy is given by:

$$S\_A = -\text{Tr}(\rho\_A \log \rho\_A)$$

Here, ρ _A is the reduced density matrix of subsystem A, obtained by tracing out the degrees of freedom of subsystem B from the joint density matrix of the entire system. TR denotes the trace operation, and log represents the logarithm with base 2 or natural logarithm.

The von Neumann entropy captures the amount of entanglement between subsystems A and the rest of the system. A larger entanglement entropy implies a higher degree of entanglement, indicating that there is more information shared between the subsystems.

Entanglement entropy has several important properties and applications:

1. Entanglement Spectrum: The eigenvalues of the reduced density matrix $\rho \rho_A$, known as the entanglement spectrum, provide information about the energy distribution and correlations in the quantum system. The entanglement spectrum can be used to characterize different phases of matter and detect quantum phase transitions.

2. Quantum Field Theory: Entanglement entropy

has been extensively studied in the context of quantum field theory, particularly in the framework of the AdS/CFT correspondence. It has provided insights into the holographic nature of entanglement and the relationship between entanglement entropy and black hole physics.

3. Quantum Criticality: Near quantum phase transitions, the entanglement entropy exhibits scaling behaviour, which can be used to identify critical points and classify different universality classes of phase transitions.

4. Quantum Information Processing: Entanglement entropy plays a crucial role in quantum information processing tasks such as quantum state tomography, quantum error correction, and quantum channel capacity calculations. It provides a measure of the resources required for various quantum information protocols.

Understanding and quantifying entanglement entropy is essential for studying complex quantum systems, including quantum many-body systems and quantum information processing. It offers valuable insights into the nature of quantum entanglement and its role in various physical phenomena.

What are Quantum Experiment Tests of Foundation?

Quantum experiments and tests of foundations refer to various scientific investigations aimed at understanding the fundamental principles and properties of quantum mechanics. These experiments aim to probe the counterintuitive nature of quantum theory and verify its predictions through rigorous experimental setups. Here are a few notable quantum experiments and tests of foundations:

1. Double-Slit Experiment: The double-slit experiment is a classic demonstration of the wave-particle duality of quantum particles, such as electrons or photons. It involves shining a

beam of particles through two parallel slits, which creates an interference pattern on a screen, suggesting that particles exhibit wave-like behaviour.

2. Bell's Theorem and Bell Tests: Bell's theorem is a fundamental result in quantum mechanics that establishes the incompatibility between certain local hidden variable theories and the predictions of quantum theory. Bell tests are experimental setups designed to test the violation of Bell's inequalities, which provide a quantitative way to distinguish between classical and quantum correlations.

3. Quantum Entanglement Experiments: Quantum entanglement is a phenomenon where two or more particles become correlated in such a way that the state of one particle cannot be described independently of the others. Experiments, such as the Aspect experiment,

have been conducted to confirm the violation of Bell's inequalities and demonstrate the non-local correlations arising from entanglement.

4. Delayed Choice Quantum Eraser: This experiment explores the wave-particle duality and the role of measurement in quantum systems. It involves the manipulation of entangled photons and the ability to make "delayed choices" about how to measure their properties, demonstrating retroactive effects on the particles behaviour.

5. Quantum Teleportation: Quantum teleportation is a process by which the exact state of a quantum system can be transmitted from one location to another. Experiments have successfully demonstrated the teleportation of quantum states using entangled particles.

6. Quantum Computing Experiments: Quantum computing experiments focus on utilizing

quantum systems to perform computational tasks that are beyond the capabilities of classical computers. These experiments involve manipulating and measuring quantum bits (qubits) to perform quantum algorithms and solve specific problems.

These are just a few examples of the numerous experiments and tests conducted to study the foundations of quantum mechanics. Quantum theory is a thriving area of research, and scientists continue to explore its intriguing properties and potential applications.

END OF FIRST SERIES

Acknowledgements

Thanks To all of my Family and Friends who have supported me to write this wonderful book.

The websites used for this books are listed down

Wikipedia

O'Reilly

Geeks for Geeks

YouTube for understanding this concepts

SciSpace

Harvard Review Papers

Written By:

BHAGAVATULA PRANAV SRINIVAS